Contents

L1B CONTENTS

Use of guidance

THE APPROVED DOCUMENTS

This document is one of a series that has been approved and issued by the Secretary of State for the purpose of providing practical guidance with respect to the technical requirements of the Building Regulations 2000 for England and Wales.

At the back of this document is a list of all the documents that have been approved and issued by the Secretary of State for this purpose.

Approved Documents are intended to provide guidance for some of the more common building situations. However, there may well be alternative ways of achieving compliance with the requirements. Thus there is no obligation to adopt any particular solution contained in an Approved Document if you prefer to meet the relevant requirement in some other way.

OTHER REQUIREMENTS

The guidance contained in an Approved Document relates only to the particular requirements of the Regulations that the document addresses. The building work will also have to comply with the requirements of any other relevant paragraphs in Schedule 1 to the Regulations.

There are Approved Documents that give guidance on each of the Parts of Schedule 1 and on Regulation 7.

LIMITATION ON REQUIREMENTS

In accordance with Regulation 8, the requirements in Parts A to D, F to K and N (except for paragraphs H2 and J6 of Schedule 1 to the Building Regulations) do not require anything to be done except for the purpose of securing reasonable standards of health and safety for persons in or about buildings (and any others who may be affected by buildings or matters connected with buildings). This is one of the categories of purpose for which Building Regulations may be made.

Paragraphs H2 and J6 are excluded from Regulation 8 because they deal directly with prevention of the contamination of water. Parts E and M (which deal, respectively, with resistance to the passage of sound, and access to and use of buildings) are excluded from Regulation 8 because they address the welfare and convenience of building users. Part L is excluded from Regulation 8 because it addresses the conservation of fuel and power. All these matters are amongst the purposes, other than health and safety, that may be addressed by Building Regulations.

MATERIALS AND WORKMANSHIP

Any building work which is subject to the requirements imposed by Schedule 1 to the Building Regulations should, in accordance with Regulation 7, be carried out with proper materials and in a workmanlike manner.

You may show that you have complied with Regulation 7 in a number of ways. These include the appropriate use of a product bearing CE marking in accordance with the Construction Products Directive (89/106/EEC)[1], the Low Voltage Directive (73/23/EEC and amendment 93/68/EEC)[2] and the EMC Directive (89/336/EEC)[3] as amended by the CE Marking Directive (93/68/EEC)[4] or a product complying with an appropriate technical specification (as defined in those Directives), a British Standard, or an alternative national technical specification of any state which is a contracting party to the European Economic Area which, in use, is equivalent, or a product covered by a national or European certificate issued by a European Technical Approval Issuing body, and the conditions of use are in accordance with the terms of the certificate. You will find further guidance in the Approved Document supporting Regulation 7 on materials and workmanship.

INDEPENDENT CERTIFICATION SCHEMES

There are many UK product certification schemes. Such schemes certify compliance with the requirements of a recognised document that is appropriate to the purpose for which the material is to be used. Materials which are not so certified may still conform to a relevant standard.

Many certification bodies that approve such schemes are accredited by UKAS.

TECHNICAL SPECIFICATIONS

Building Regulations are made for specific purposes: health and safety, energy conservation and the welfare and convenience of disabled people. Standards and technical approvals are relevant guidance to the extent that they relate to these considerations. However, they may also address other aspects of performance such as serviceability, or aspects which although they relate to health and safety are not covered by the Regulations.

---

[1] As implemented by the Construction Products Regulations 1991 (SI 1991/1620).

[2] As implemented by the Electrical Equipment (Safety) Regulations 1994 (SI 1994/3260).

[3] As implemented by the Electromagnetic Compatibility Regulations 1992 (SI 1992/2372).

[4] As implemented by the Construction Products (Amendment) Regulations 1994 (SI 1994/3051) and the Electromagnetic Compatibility (Amendment) Regulations 1994 (SI 1994/3080).

When an Approved Document makes reference to a named standard, the relevant version of the standard is the one listed at the end of the publication. However, if this version has been revised or updated by the issuing standards body, the new version may be used as a source of guidance provided it continues to address the relevant requirements of the Regulations.

The appropriate use of a product that complies with a European Technical Approval as defined in the Construction Products Directive will meet the relevant requirements.

The Office intends to issue periodic amendments to its Approved Documents to reflect emerging harmonised European Standards. Where a national standard is to be replaced by a European harmonised standard, there will be a coexistence period during which either standard may be referred to. At the end of the coexistence period the national standard will be withdrawn.

THE WORKPLACE (HEALTH, SAFETY AND WELFARE) REGULATIONS 1992

The Workplace (Health, Safety and Welfare) Regulations 1992 as amended by The Health and Safety (Miscellaneous Amendments) Regulations 2002 (SI 2002/2174) contain some requirements which affect building design. The main requirements are now covered by the Building Regulations, but for further information see: *Workplace health, safety and welfare: Workplace (Health, Safety and Welfare) Regulations 1992, Approved Code of Practice,* L24, HMSO, 1992 (ISBN 0 71760 413 6).

The Workplace (Health, Safety and Welfare) Regulations 1992 apply to the common parts of flats and similar buildings if people such as cleaners and caretakers are employed to work in these common parts. Where the requirements of the Building Regulations that are covered by this Part do not apply to dwellings, the provisions may still be required in the situations described above in order to satisfy the Workplace Regulations.

MIXED USE DEVELOPMENT

In mixed use developments part of a building may be used as a dwelling while another part has a non-domestic use. In such cases, if the requirements of this Part of the Regulations for dwellings and non-domestic use differ, the requirements for non-domestic use should apply in any shared parts of the building.

The Requirement

This Approved Document, which takes effect on 6 April 2006, deals with the energy efficiency requirements in the Building Regulations 2000 (as amended by SI 2001/3335 and SI 2006/652).

The energy efficiency requirements are conveyed in Part L of Schedule 1 to the Regulations and regulations 4A, 17C and 17D as described below.

Requirement	*Limits on application*

Part L Conservation of fuel and power
 	
L1. Reasonable provision shall be made for the conservation of fuel and power in buildings by:
 	
   a. limiting heat gains and losses:
 	
      i. through thermal elements and other parts of the building fabric; and
 	
      ii. from pipes, ducts and vessels used for space heating, space cooling and hot water services;
 	
   b. providing and commissioning energy efficient fixed building services with effective controls; and
 	
   c. providing to the owner sufficient information about the building, the fixed building services and their maintenance requirements so that the building can be operated in such a manner as to use no more fuel and power than is reasonable in the circumstances.	

Other changes to the Regulations

There are new Regulations that introduce new energy efficiency requirements and other relevant changes to the existing regulations.

For ease of reference the principal elements of the regulations that bear on energy efficiency are repeated below and, where relevant, in the body of the guidance in the rest of this Approved Document. However it must be recognised that the Statutory Instrument takes precedence if there is any doubt over interpretation.

Interpretation

Regulation 2(1) is amended to include the following new definitions.

'Change to a building's energy status' means any change which results in a building becoming a building to which the energy efficiency requirements of these Regulations apply, where previously it was not.

'Energy efficiency requirements' means the requirements of regulations 4A, 17C and 17D and Part L of Schedule 1.

'Fixed building services' means any part of, or any controls associated with:

a. fixed internal or external lighting systems, but does not include emergency escape lighting or specialist process lighting; or

b. fixed systems for heating, hot water service, air conditioning or mechanical ventilation.

'Renovation' in relation to a thermal element means the provision of a new layer in the thermal element or the replacement of an existing layer, but excludes decorative finishes, and 'renovate' shall be construed accordingly.

New paragraphs (2A) and (2B) are added to Regulation 2 as follows.

(2A) In these 'thermal element' means a wall, floor or roof (but does not include windows, doors, roof windows or roof-lights) which separates a thermally conditioned part of the building ('the conditioned space') from:

a. the external environment (including the ground); or

b. in the case of floors and walls, another part of the building which is:

   i. unconditioned;

   ii. an extension falling within class VII in Schedule 2; or

   iii. where this paragraph applies, conditioned to a different temperature,

and includes all parts of the element between the surface bounding the conditioned space and the external environment or other part of the building as the case may be.

(2B) Paragraph (2A)(b)(iii) only applies to a building which is not a dwelling, where the other part of the building is used for a purpose which is not similar or identical to the purpose for which the conditioned space is used.

Meaning of building work

Regulation 3 is amended as follows.

3.–(1) In these Regulations 'building work' means:

a. the erection or extension of a building;

b. the provision or extension of a controlled service or fitting in or in connection with a building;

c. the material alteration of a building, or a controlled service or fitting, as mentioned in paragraph (2);

d. work required by regulation 6 (requirements relating to material change of use);

e. the insertion of insulating material into the cavity wall of a building;

f. work involving the underpinning of a building;

g. work required by regulation 4A (requirements relating to thermal elements);

h. work required by regulation 4B (requirements relating to a change of energy status);

i. work required by regulation 17D (consequential improvements to energy performance).

(2) An alteration is material for the purposes of these Regulations if the work, or any part of it, would at any stage result:

a. in a building or controlled service or fitting not complying with a relevant requirement where previously it did; or

b. in a building or controlled service or fitting which before the work commenced did not comply with a relevant requirement, being more unsatisfactory in relation to such a requirement.

(3) In paragraph (2) 'relevant requirement' means any of the following applicable requirements of Schedule 1, namely:

   Part A (structure)
   paragraph B1 (means of warning and escape)
   paragraph B3 (internal fire spread – structure)
   paragraph B4 (external fire spread)
   paragraph B5 (access and facilities for the fire service)
   Part M (access to and use of buildings).

Requirements relating to building work

Regulation 4 is amended as follows.

4.–(1) Subject to paragraph 1A building work shall be carried out so that:

  a. it complies with the applicable requirements contained in Schedule 1; and

  b. in complying with any such requirement there is no failure to comply with any other such requirement.

(1A) Where:

  a. building work is of a kind described in regulation 3(1)(g), (h) or (i); and

  b. the carrying out of that work does not constitute a material alteration,

that work need only comply with the applicable requirements of Part L of Schedule 1.

(2) Building work shall be carried out so that, after it has been completed:

  a. any building which is extended or to which a material alteration is made; or

  b. any building in, or in connection with, which a controlled service or fitting is provided, extended or materially altered; or

  c. any controlled service or fitting,

complies with the applicable requirements of Schedule 1 or, where it did not comply with any such requirement, is no more unsatisfactory in relation to that requirement than before the work was carried out.

Requirements relating to thermal elements

A new regulation 4A is added as follows.

4A.–(1) Where a person intends to renovate a thermal element, such work shall be carried out as is necessary to ensure that the whole thermal element complies with the requirements of paragraph L1(a)(i) of Schedule 1.

(2) Where a thermal element is replaced, the new thermal element shall comply with the requirements of paragraph L1(a)(i) of Schedule 1.

Requirements relating to a change to energy status

A new regulation 4B is added as follows.

4B.–(1) Where there is a change to a building's energy status, such work, if any, shall be carried out as is necessary to ensure that the building complies with the applicable requirements of Part L of Schedule 1.

(2) In this regulation 'building' means the building as a whole or parts of it that have been designed or altered to be used separately.

Requirements relating to a material change of use

Regulation 6 is updated to take account of the changes to Part L.

Exempt buildings and work

Regulation 9 is substantially altered as follows.

9.–(1) Subject to paragraphs (2) and (3) these Regulations do not apply to:

  a. the erection of any building or extension of a kind described in Schedule 2; or

  b. the carrying out of any work to or in connection with such a building or extension, if after the carrying out of that work it is still a building or extension of a kind described in that Schedule.

(2) The requirements of Part P of Schedule 1 apply to:

  a. any greenhouse;

  b. any small detached building falling within class VI in Schedule 2; and

  c. any extension of a building falling within class VII in Schedule 2,

which in any case receives its electricity from a source shared with or located inside a dwelling.

(3) The energy efficiency requirements of these Regulations apply to:

  a. the erection of any building of a kind falling within this paragraph;

  b. the extension of any such building, other than an extension falling within class VII in Schedule 2; and

  c. the carrying out of any work to or in connection with any such building or extension.

(4) A building falls within paragraph (3) if it:

  a. is a roofed construction having walls;

  b. uses energy to condition the indoor climate; and

  c. does not fall within the categories listed in paragraph (5).

(5) The categories referred to in paragraph (4)(c) are:

  a. buildings which are:

    i. listed in accordance with section 1 of the Planning (Listed Buildings and Conservation Areas) Act 1990;

    ii. in a conservation area designated in accordance with section 69 of that Act; or

    iii. included in the schedule of monuments maintained under section 1 of the Ancient Monuments and Archaeological Areas Act 1979,

where compliance with the energy efficiency requirements would unacceptably alter their character or appearance;

b. buildings which are used primarily or solely as places of worship;

c. temporary buildings with a planned time of use of two years or less, industrial sites, workshops and non-residential agricultural buildings with low energy demand;

d. stand-alone buildings other than dwellings with a total useful floor area of less than 50m².

(6) In this regulation:

a. 'building' means the building as a whole or parts of it that have been designed or altered to be used separately; and

b. the following terms have the same meaning as in European Parliament and Council Directive 2002/91/EC on the energy performance of buildings:

   i.   'industrial sites';

   ii. 'low energy demand';

   iii. 'non-residential agricultural buildings';

   iv. 'places of worship';

   v.   'stand alone';

   vi. 'total useful floor area';

   vii. 'workshops'.

Giving of a building notice or deposit of plans

Regulation 12 is substantially amended as follows.

12.–(1) In this regulation 'relevant use' means a use as a workplace of a kind to which Part II of the Fire Precautions (Workplace) Regulations 1997 applies or a use designated under section 1 of the Fire Precautions Act 1971.

(2) This regulation applies to a person who intends to:

a. carry out building work;

b. replace or renovate a thermal element in a building to which the energy efficiency requirements apply;

c. make a change to a building's energy status; or

d. make a material change of use.

(2A) Subject to the following provisions of this regulation, a person to whom this regulation applies shall:

a. give to the local authority a building notice in accordance with regulation 13; or

b. deposit full plans with the local authority in accordance with regulation 14.

(3) A person shall deposit full plans where he intends to carry out building work in relation to a building to which the Regulatory Reform (Fire Safety) Order 2005 applies, or will apply after the completion of the building work.

(4) A person shall deposit full plans where he intends to carry out work which includes the erection of a building fronting on to a private street.

(4A) A person shall deposit full plans where he intends to carry out building work in relation to which paragraph H4 of Schedule 1 imposes a requirement.

(5) A person who intends to carry out building work is not required to give a building notice or deposit full plans where the work consists only of work:

a. described in column 1 of the Table in Schedule 2A if the work is to be carried out by a person described in the corresponding entry in column 2 of that Table, and paragraphs 1 and 2 of that schedule have effect for the purposes of the descriptions in the Table; or

b. described in Schedule 2B.

(6) Where regulation 20 of the Building (Approved Inspectors etc.) Regulations 2000 (local authority powers in relation to partly completed work) applies, the owner shall comply with the requirements of that regulation instead of with this regulation.

(7) Where:

a. a person proposes to carry out work which consists of emergency repairs to any fixed building services in respect of which Part L of Schedule 1 imposes a requirement;

b. it is not practicable to comply with paragraph (2A) before commencing the work; and

c. paragraph (5) does not apply,

he shall give a building notice to the local authority as soon as reasonably practicable after commencement of the work.

Regulation 13 (particulars and plans where a building notice is given) and 14 (full plans)

These are amended to apply to renovation or replacement of a thermal element and a change to a building's energy status.

Provisions applicable to self-certification schemes

16A.–(1) This regulation applies to the extent that the building work consists only of work of a type described in column 1 of the Table in Schedule 2A and the work is carried out by a person who is described in the corresponding entry in column 2 of that Table in respect of that type of work.

(2) Where this regulation applies, the local authority is authorised to accept, as evidence that the requirements of regulations 4 and 7 have been satisfied, a certificate to that effect by the person carrying out the building work.

(3) Where this regulation applies, the person carrying out the work shall, not more than 30 days after the completion of the work -

(a) give to the occupier a copy of the certificate referred to in paragraph (2); and

(b) give to the local authority -

    (i) notice to that effect, or

    (ii) the certificate referred to in paragraph (2).

(4) Paragraph (3) of this regulation does not apply where a person carries out the building work described in Schedule 2B.

New Part VA

Energy performance of buildings

New Regulations are added as follows.

Methodology of calculation of the energy performance of buildings

17A. The Secretary of State shall approve a methodology of calculation of the energy performance of buildings.

Minimum energy performance requirements for buildings

17B. The Secretary of State shall approve minimum energy performance requirements for new buildings, in the form of target CO_2 emission rates, which shall be based upon the methodology approved pursuant to regulation 17A.

New buildings

17C. Where a building is erected, it shall not exceed the target CO_2 emission rate for the building that has been approved pursuant to regulation 17B.

Consequential improvements to energy performance

17D.–(1) Paragraph (2) applies to an existing building with a total useful floor area over 1000m^2 where the proposed building work consists of or includes:

a. an extension;

b. the initial provision of any fixed building services; or

c. an increase to the installed capacity of any fixed building services.

(2) Subject to paragraph (3), where this regulation applies, such work, if any, shall be carried out as is necessary to ensure that the building complies with the requirements of Part L of Schedule 1.

(3) Nothing in paragraph (2) requires work to be carried out if it is not technically, functionally and economically feasible.

Interpretation

17E. In this Part 'building' means the building as a whole or parts of it that have been designed or altered to be used separately.

Part VI – Miscellaneous

New Regulations are added as follows.

Pressure testing

20B.–(1) This regulation applies to the erection of a building in relation to which paragraph L1(a)(i) of Schedule 1 imposes a requirement.

(2) Where this regulation applies, the person carrying out the work shall, for the purpose of ensuring compliance with regulation 17C and paragraph L1(a)(i) of Schedule 1:

a. ensure that:

  i.   pressure testing is carried out in such circumstances as are approved by the Secretary of State; and

  ii. the testing is carried out in accordance with a procedure approved by the Secretary of State; and

b. subject to paragraph (5), give notice of the results of the testing to the local authority.

(3) The notice referred to in paragraph (2)(b) shall:

a. record the results and the data upon which they are based in a manner approved by the Secretary of State; and

b. be given to the local authority not later than seven days after the final test is carried out.

(4) A local authority is authorised to accept, as evidence that the requirements of paragraph (2)(a)(ii) have been satisfied, a certificate to that effect by a person who is registered by the British Institute of Non-destructive Testing in respect of pressure testing for the airtightness of buildings.

(5) Where such a certificate contains the information required by paragraph (3)(a), paragraph (2)(b) does not apply.

Commissioning

20C.–(1) This regulation applies to building work in relation to which paragraph L1(b) of Schedule 1 imposes a requirement, but does not apply where the work consists only of work described in Schedule 2B.

(2) Where this regulation applies the person carrying out the work shall, for the purpose of ensuring compliance with paragraph L1(b) of Schedule 1, give to the local authority a notice confirming that the fixed building services have been commissioned in accordance with a procedure approved by the Secretary of State.

(3) The notice shall be given to the local authority:

a. not later than the date on which the notice required by regulation 15(4) is required to be given; or

b. where that regulation does not apply, not more than 30 days after completion of the work.

CO_2 emission rate calculations

20D.–(1) Subject to paragraph (4), where regulation 17C applies the person carrying out the work shall provide to the local authority a notice which specifies:

a. the target CO_2 emission rate for the building; and

b. the calculated CO_2 emission rate for the building as constructed.

(2) The notice shall be given to the local authority not later than the date on which the notice required by regulation 20B is required to be given.

(3) A local authority is authorised to accept, as evidence that the requirements of regulation 17C would be satisfied if the building were constructed in accordance with an accompanying list of specifications, a certificate to that effect by a person who is registered by:

a. FAERO Limited; or

b. BRE Certification Limited,

in respect of the calculation of CO_2 emission rates of buildings.

(4) Where such a certificate is given to the local authority:

a. paragraph (1) does not apply; and

b. the person carrying out the work shall provide to the local authority not later than the date on which the notice required by regulation 20B is required to be given a notice which:

  i.   states whether the building has been constructed in accordance with the list of specifications which accompanied the certificate; and

  ii. if it has not, lists any changes to the specifications to which the building has been constructed.

Schedule 2A

Schedule 2A is amended as follows:

Self-certification schemes and exemptions from requirement to give building notice or deposit full plans.

Column 1	Column 2
Type of work	*Person carrying out work*
1. Installation of a heat-producing gas appliance.	A person, or an employee of a person, who is a member of a class of persons approved in accordance with regulation 3 of the Gas Safety (Installation and Use) Regulations 1998.
2. Installation of heating or hot water service system connected to a heat-producing gas appliance, or associated controls.	A person registered by CORGI Services Limited in respect of that type of work.
3. Installation of: a. an oil-fired combustion appliance which has a rated heat output of 100 kilowatts or less and which is installed in a building with no more than 3 storeys (excluding any basement) or in a dwelling; b. oil storage tanks and the pipes connecting them to combustion appliances; or c. heating and hot water service systems connected to an oil-fired combustion appliance.	An individual registered by Oil Firing Technical Association Limited, NAPIT Certification Limited or Building Engineering Services Competence Accreditation Limited in respect of that type of work.
4. Installation of: a. a solid fuel burning combustion appliance which has a rated heat output of 50 kilowatts or less which is installed in a building with no more than 3 storeys (excluding any basement); or b. heating and hot water service systems connected to a solid fuel burning combustion appliance.	A person registered by HETAS Limited, NAPIT Certification Limited or Building Engineering Services Competence Accreditation Limited in respect of that type of work.
5. Installation of a heating or hot water service system, or associated controls, in a dwelling.	A person registered by Building Engineering Services Competence Accreditation Limited in respect of that type of work.
6. Installation of a heating, hot water service, mechanical ventilation or air conditioning system, or associated controls, in a building other than a dwelling.	A person registered by Building Engineering Services Competence Accreditation Limited in respect of that type of work.
7. Installation of an air conditioning or ventilation system in an existing dwelling, which does not involve work on systems shared with other dwellings.	A person registered by CORGI Services Limited or NAPIT Certification Limited in respect of that type of work.
8. Installation of a commercial kitchen ventilation system which does not involve work on systems shared with parts of the building occupied separately.	A person registered by CORGI Services Limited in respect of that type of work.
9. Installation of a lighting system or electric heating system, or associated electrical controls.	A person registered by The Electrical Contractors Association Limited in respect of that type of work.
10. Installation of fixed low or extra-low voltage electrical installations.	A person registered by BRE Certification Limited, British Standards Institution, ELECSA Limited, NICEIC Group Limited or NAPIT Certification Limited in respect of that type of work.
11. Installation of fixed low or extra-low voltage electrical installations as a necessary adjunct to or arising out of other work being carried out by the registered person.	A person registered by CORGI Services Limited, ELECSA Limited, NAPIT Certification Limited, NICEIC Group Limited or Oil Firing Technical Association Limited in respect of that type of electrical work.
12. Installation, as a replacement, of a window, rooflight, roof window or door (being a door which together with its frame has more than 50 per cent of its internal face area glazed) in an existing building.	A person registered under the Fenestration Self-Assessment Scheme by Fensa Ltd, or by CERTASS Limited or the British Standards Institution in respect of that type of work.

13. Installation of a sanitary convenience, washing facility or bathroom in a dwelling, which does not involve work on shared or underground drainage.	A person registered by CORGI Services Limited or NAPIT Certification Limited in respect of that type of work.

14.–(1) Subject to paragraph (2), any building work, other than the provision of a masonry chimney, which is necessary to ensure that any appliance, service or fitting which is installed and which is described in the preceding entries in column 1 above, complies with the applicable requirements contained in Schedule 1.
 	
(2) Paragraph (1) does not apply to:
 	
  c. building work which is necessary to ensure that a heat-producing gas appliance complies with the applicable requirements contained in Schedule 1 unless the appliance:
 	
     i. has a rated heat output of 100 kilowatts or less; and
 	
     ii. is installed in a building with no more than 3 storeys (excluding any basement), or in a dwelling; or
 	
  d. the provision of a masonry chimney.
	The person who installs the appliance, service or fitting to which the building work relates and who is described in the corresponding entry in column 2 above.

Schedule 2B

Schedule 2B is amended as follows.

Descriptions of work where no building notice or deposit of full plans required.

1 Work consisting of:

a. replacing any fixed electrical equipment which does not include the provision of:

    i.   any new fixed cabling;

    ii. a new consumer unit; and

b. replacing a damaged cable for a single circuit only;

c. re-fixing or replacing enclosures of existing installation components, where the circuit protective measures are unaffected;

d. providing mechanical protection to an existing fixed installation, where the circuit protective measures and current carrying capacity of conductors are unaffected by the increased thermal insulation;

e. installing or upgrading main or supplementary equipotential bonding;

f. in heating or cooling systems:

    i.   replacing control devices that utilise existing fixed control wiring or pneumatic pipes;

    ii. replacing a distribution system output device;

    iii. providing a valve or a pump;

    iv. providing a damper or fan;

g. in hot water service systems, providing a valve or pump;

h. replacing an external door (where the door together with its frame has not more than 50% of its internal face area glazed);

i. in existing buildings other than dwellings, providing fixed internal lighting where no more than 100m^2 of the floor area of the building is to be served by the lighting.

2 Work which:

a. is not in a kitchen, or a special location;

b. does not involve work on a special installation; and

c. consists of:

    i.   adding light fittings and switches to an existing circuit; or

    ii. adding socket outlets and fused spurs to an existing ring or radial circuit.

3 Work on:

a. telephone wiring or extra-low voltage wiring for the purposes of communications, information technology, signalling, control and similar purposes, where the wiring is not in a special location;

b. equipment associated with the wiring referred to in sub-paragraph (a).

c. pre-fabricated equipment sets and associated flexible leads with integral plug and socket connections.

4 For the purposes of this Schedule:

'kitchen' means a room or part of a room which contains a sink and food preparation facilities;

'special installation' means an electric floor or ceiling heating system, an outdoor lighting or electric power installation, an electricity generator, or an extra-low voltage lighting system which is not a pre-assembled lighting set bearing the CE marking referred to in regulation 9 of the Electrical Equipment (Safety) Regulations 1994; and

'special location' means a location within the limits of the relevant zones specified for a bath, a shower, a swimming or paddling pool or a hot air sauna in the Wiring Regulations, sixteenth edition, published by the Institution of Electrical Engineers and the British Standards Institution as BS 7671: 2001 and incorporating amendments 1 and 2.

Section 0: General guidance

CONVENTIONS USED IN THIS DOCUMENT

1 In this document the following conventions have been adopted to assist understanding and interpretation:

a. Texts shown against a green background are extracts from the Building Regulations as amended and convey the legal requirements that bear on compliance with Part L. It should be remembered however that building works must comply with all the other relevant provisions. Similar provisions are conveyed by the Building (Approved Inspectors) Regulations as amended.

b. Key terms are printed in **bold italic text** and defined for the purposes of this Approved Document in Section 5 of this document.

c. References given as footnotes and repeated as end notes are given as ways of meeting the requirements or as sources of more general information as indicated in the particular case. The Approved Document will be amended from time to time to include new references and to refer to revised editions where this aids compliance.

d. Additional *commentary in italic text* appears after some numbered paragraphs. The commentary is intended to assist understanding of the immediately preceding paragraph or sub-paragraph, but is not part of the approved guidance.

TYPES OF WORK COVERED BY THIS APPROVED DOCUMENT

2 This Approved Document gives guidance on what, in ordinary circumstances, will meet the requirements of Regulation 4A and Part L when carrying out different classes of building work on existing **dwellings**.

3 In particular guidance is given on the following activities:

a. extensions (see paragraphs 14 to 24)

b. when creating a new **dwelling** or part of a dwelling through a material change of use (paragraphs 25 to 28)

c. material alterations to existing **dwellings** (paragraphs 29 to 30)

d. the provision of a controlled fitting (paragraphs 32 to 34)

e. the provision or extension of a controlled service (paragraphs 35 to 48)

f. the provision or **renovation** of a **thermal element** (paragraphs 49 to 57).

4 Where the activities include building work in a **dwelling** that is part of a mixed use building, account should also be taken of the guidance in Approved Document L2B in relation to those parts of the building that are not **dwellings**, including any common areas.

It should be noted that dwellings refer to self-contained units. Rooms for residential purposes are not dwellings, and so Approved Document L2B applies to them.

5 The **energy efficiency requirements**, *apart from those in regulation 17C and 17D*, apply to work in existing dwellings. In most instances, this will require the **BCB** to be notified of the intended work before the work commences, either in the form of a deposit of full plans or by a building notice. In certain situations however other procedures apply. These include:

a. Where the work is being carried out under the terms of an approved Competent Persons self-certification (CP) scheme. In these cases, in accordance with Regulation 16A and Schedule 2A[5] no advance notification to the building control body is needed. At the completion of the work, the registered CP provides the building owner with a certificate confirming that the installation has been carried out in accordance with the requirements of the relevant requirements, and the scheme operator notifies the local authority to that effect.

b. Where the work involves an emergency repair, e.g. a failed boiler or a leaking hot water cylinder. In these cases, in accordance with Regulation 12 (7)[6], there is no need to delay making the repair in order to make an advance notification to the building control body. However, in such cases it will still be necessary for the work to comply with the requirements and to give a notice to the **BCB** at the earliest opportunity, unless an installer registered under an appropriate CP scheme carries out the work. A completion certificate can then be issued in the normal way.

c. Where the work is of a minor nature as described in Schedule 2B[7] of the Building Regulations. Again, the work must comply with the relevant requirements, but need not be notified to building control.

---

[5] A copy of these can be seen on pages 8 and 12 respectively.

[6] A copy of this can be seen on page 8.

[7] A copy of this can be seen on page 14.

TECHNICAL RISK

6 Building work must satisfy all the technical requirements set out in Regulations 4A and Schedule 1 of the Building Regulations. Part B (Fire safety), Part E (Resistance to the passage of sound), Part F (Ventilation), Part C (Site preparation and resistance to moisture), Part J (Combustion appliances and fuel storage systems) and Part P (Electrical safety) are particularly relevant when considering the incorporation of energy efficiency measures.

7 The inclusion of any particular energy efficiency measure should not involve excessive technical risk. BR 262[8] provides general guidance on avoiding risks in the application of thermal insulation.

HISTORIC BUILDINGS

8 Special considerations apply if the building on which the work is to be carried out has special historic or architectural value and compliance with the *energy efficiency requirements* would unacceptably alter the character or appearance[9].

9 When undertaking work on or in connection with buildings with special historic or architectural value, the aim should be to improve energy efficiency where and to the extent that it is practically possible. This is provided that the work does not prejudice the character of the host building or increase the risk of long-term deterioration to the building fabric or fittings. The guidance given in the English Heritage publication[10] should be taken into account in determining appropriate energy performance standards for such building works. Particular issues relating to work in historic buildings that warrant sympathetic treatment and where advice from others could therefore be beneficial include:

a. restoring the historic character of a building that has been subject to previous inappropriate alteration, e.g. replacement windows, doors and rooflights;

b. rebuilding a former building (e.g. following a fire or filling a gap site in a terrace);

c. making provisions enabling the fabric of historic buildings to 'breathe' to control moisture and potential long-term decay problems.

10 In arriving at a balance between historic building conservation and energy efficiency improvements, it would be appropriate to take into account the advice of the local authority's conservation officer.

CALCULATION OF U-VALUES

11 U-values must be calculated using the methods and conventions set out in BR 443[11], 'Conventions for U-value calculations'.

12 The U-values for roof windows and rooflights given in this Approved Document are based on the U-value having been assessed with the roof window or rooflight in the vertical position. If a particular unit has been assessed in a plane other than the vertical, the standards given in this Approved Document should be modified by making a U-value adjustment following the guidance given in BR 443.

For example: the standard for a replacement rooflight in Table 2 is 2.0W/m²·K. This is for the unit assessed in the vertical plane. The performance of a double-glazed rooflight in the horizontal plane, based on the guidance given in BR 443, would be adjusted by 0.5W/m²·K to 2.0 + 0.5 = 2.5W/m²·K.

BUILDINGS THAT ARE EXEMPT FROM THE REQUIREMENTS IN PART L

13 The provisions for exempting buildings and building work from the Building Regulations requirements have changed and the new provisions are given in regulation 9.

---

[8] BR 262 *Thermal insulation: avoiding risks*, BRE, 2001.

[9] See the copy of regulation 9 on page 7.

[10] *Building Regulations and Historic Buildings*, English Heritage, 2002 (revised 2004).

[11] BR 443 *Conventions for U-value calculations*, BRE, 2006.

Section 1: Guidance relating to building work

THE EXTENSION OF A DWELLING

Fabric standards

14 Reasonable provision would be for the proposed extension to achieve the following performance standards:

a. Controlled fittings that meet the standards set out in paragraphs 32 to 34 of this Approved Document.

b. Newly constructed **thermal elements** that meet the standards set out in paragraphs 49 to 53 of this Approved Document.

c. When working on existing fabric elements that are to become thermal elements a way of complying would be to follow the guidance in paragraphs 54 and 57.

Area of windows, roof windows and doors

15 In most circumstances reasonable provision would be to limit the area of windows, roof windows and doors in extensions so that it does not exceed the sum of:

a. 25% of the floor area of the extension; plus

b. the area of any windows or doors which, as a result of the extension works, no longer exist or are no longer exposed.

16 In some cases different approaches may be adopted by agreement with the **BCB** in order to achieve a satisfactory level of daylighting. BS 8206[12] gives guidance on this.

Heating and lighting in the extension

17 Where a **fixed building service** is provided or extended as part of constructing the extension, reasonable provision would be to follow the guidance in paragraphs 35 to 48.

Optional approaches with more design flexibility

18 More flexibility in the selection of U-values and opening areas than is available by following the guidance in paragraphs 14 and 15 can be obtained by compensation elsewhere in the design. A way of complying would be to show that:

a. the area-weighted U-value of all the elements in the extension is no greater than that of an extension of the same size and shape that complies with the U-value standards referred to in paragraph 14 and the opening area in paragraph 15; and

The area-weighted U-value is given by the following expression:

$$\{(U_1 \times A_1) + (U_2 \times A_2) + (U_3 \times A_3) + \ldots)\} \div \{(A_1 + A_2 + A_3 + \ldots)\}$$

b. the area-weighted U-value for each element type is no worse than the value given in column (a) of Table 1; and

c. the U-value of any individual element is no worse than the relevant value in column (b) of Table 1.

To minimise condensation risk in localised parts of the envelope. Individual elements are defined as those areas of the given element type that have the same construction details.

19 In the case of windows, doors and rooflights, the assessment should be based on the whole unit (i.e. in the case of a window, the combined performance of the glazing and frame).

Table 1 **Limiting U-value standards (W/m²·K)**		
Element	(a) Area-weighted average U-value	(b) Limiting U-value
Wall	0.35	0.70
Floor	0.25	0.70
Roof	0.25	0.35
Windows, roof windows, rooflights[1] and doors	2.2	3.3

Notes:

1 See paragraph 12.

20 Where even greater design flexibility is required, reasonable provision would be to use SAP 2005[13] to show that the calculated CO_2 emission rate from the **dwelling** with its proposed extension is no greater than for the **dwelling** plus a notional extension built to the standards of paragraphs 14 to 17. In these cases the area-weighted average U-value of each element type should be no worse than the standards set out in column (a) of Table 1, and the U-value of any individual element should be no worse than the values in column (b) of Table 1. The data in SAP 2005 Appendix S can be used to estimate the performance of the elements of the existing building where these are unknown.

21 If, as part of achieving the standard set out in paragraph 20, improvements are proposed to the existing **dwelling**, such improvements should be implemented to a standard that is no worse than set out in the relevant guidance contained in this Approved Document. The relevant standards for improving retained thermal elements are as set out in column (b) of Table 5.

---

[12] BS 8206-2:1992 Lighting for buildings. Code of practice for daylighting.

[13] The Government's Standard Assessment Procedure for Energy Rating of Dwellings, 2005 edition, SAP 2005, Defra.

Where it is proposed to upgrade, then the standards set out in this Approved Document are cost effective and should be implemented in full. It will be worthwhile implementing them even if the improvement is more than necessary to achieve compliance. In some cases therefore, the standard of the extended house may be better than that required by paragraph 20 alone. Paragraph 21 ensures that no cost-effective improvement opportunities are traded away.

Conservatories and substantially glazed extensions

22 Where the extension is a *conservatory* that is not exempt by Regulation 9(3)[14], then reasonable provision would be to provide:

a. Effective thermal separation from the heated area in the existing *dwelling*. The walls, doors and windows between the *dwelling* and the extension should be insulated and draught-stripped to at least the same extent as in the existing *dwelling*.

If a highly glazed extension is not thermally separated from the dwelling, then it should be regarded as a conventional extension. Compliance in such cases could be demonstrated using the approach set out in paragraphs 14 to 21.

b. Independent temperature and on/off controls to any heating system. Any heating appliance should also conform to the standards set out in paragraph 35.

c. Glazed elements should comply with the standards given in column (b) of Table 2 and any thermal elements should have U-values that are no worse than the standards given in column (b) of Table 4.

23 Conservatories built at ground level and with a floor area no greater than 30m² are exempt from the Building Regulations (other than having to satisfy the requirements of Part N).

24 If a substantially glazed extension fails to qualify as a *conservatory* because it has less than the minimum qualifying amounts of translucent material, but otherwise satisfies paragraph 22, reasonable provision would be to demonstrate that the performance is no worse than a *conservatory* of the same size and shape. A way of doing so would be to show the area-weighted U-value of the elements in the proposed extension is no greater than that of a *conservatory* that complies with the standards set out in paragraph 22.

MATERIAL CHANGE OF USE

25 Material changes of use involving dwellings are defined in Regulation 5 as follows:

For the purposes of paragraph 8(1)(e) of Schedule 1 to the Act and for the purposes of these Regulations, there is a material change of use where there is a change in the purposes for which or the circumstances in which a building is used, so that after that change:

a. the building is used as a dwelling, where previously it was not;

b. the building contains a flat, where previously it did not;

c. the building is used as an hotel or a boarding house, where previously it was not;

d. the building is used as an institution, where previously it was not;

e. the building is used as a public building, where previously it was not;

f. the building is not a building described in Classes I to VI in Schedule 2, where previously it was;

g. the building, which contains at least one dwelling, contains a greater or lesser number of dwellings than it did previously;

h. the building contains a room for residential purposes, where previously it did not;

i. the building, which contains at least one room for residential purposes, contains a greater or lesser number of such rooms than it did previously; or

j. the building is used as a shop where previously it was not.

26 When carrying out a material change of use, the Reasonable provision would be:

a. when carrying out a material change of use; or

b. when a building changes its energy status

to follow the guidance in paragraph 27.

27 In normal circumstances, reasonable provision would be:

a. Where controlled services or fittings are being provided or extended, to meet the standards set out in paragraphs 31 to 48 of this Approved Document.

b. Where the work involves the provision of a *thermal element*, to meet the standards set out in paragraphs 49 to 53 of this Approved Document.

For the purposes of Building Regulations, provision means both new and replacement elements.

c. Where the work involves the *renovation* of *thermal elements*, to meet the guidance in paragraphs 54 to 55 of this Approved Document.

d. Any *thermal element* that is being retained should be upgraded following the guidance given in paragraphs 56 and 57 of this Approved Document.

e. Any existing window (including roof window or rooflight) or door which separates a conditioned space from an unconditioned space or the external environment and which has a U-value that is worse than 3.3W/m²·K, should be replaced following the guidance in paragraphs 32 to 34.

---

[14] See the copy of Regulation 9 on page 7.

Option providing more design flexibility

28 To provide more design flexibility SAP 2005 can be used to demonstrate that the total CO_2 emissions from all the **dwellings** in the building as it will become are no greater than if each **dwelling** had been improved following the guidance set out in paragraph 27. In these cases the U-values of any individual element should be no worse than the values in column (b) of Table 1.

MATERIAL ALTERATION

29 Material alterations are defined in Regulation 3(2) as follows.

3(2) An alteration is material for the purposes of these Regulations if the work, or any part of it, would at any stage result:

a. in a building or controlled service or fitting not complying with a relevant requirement where previously it did; or

b. in a building or controlled service or fitting which before the work commenced did not comply with a relevant requirement, being more unsatisfactory in relation to such a requirement.'

3(3) In paragraph (2) 'relevant requirement' means any of the following applicable requirements of Schedule 1, namely:

a. Part A (structure)

b. Paragraph B1 (means of warning and escape)

c. Paragraph B3 (internal fire spread – structure)

d. Paragraph B4 (external fire spread)

e. Paragraph B5 (access and facilities for the fire service)

f. Part M (access to and use of buildings).

30 When carrying out a material alteration, reasonable provision would be

a. when the work involves the provision of a **thermal element**, to follow the guidance in paragraphs 50 to 53 of this Approved Document.

For the purposes of Building Regulations, provision means both new and replacement elements.

b. when the work involves the renovation of a **thermal element**, to follow the guidance in paragraphs 54 and 55 of this Approved Document.

c. where an existing element becomes part of the thermal envelope of the building where previously it was not, to follow the guidance in paragraphs 56 and 57 of this Approved Document.

d. when providing controlled fittings, to limit glazing area to reasonable provision and to follow the guidance on controlled fittings given in paragraphs 32 to 34 of this Approved Document.

Reasonable provision for glazing area depends on the circumstances in the particular case. A way of showing compliance would be to follow the approaches given in paragraphs 15 and 16.

e. when providing or extending a controlled service, to follow the guidance on controlled services given in paragraphs 35 to 48 of this Approved Document.

WORK ON CONTROLLED FITTINGS AND SERVICES

31 Controlled services or fittings are defined in Regulation 2 as follows:

controlled service or fitting means a service or fitting in relation to which Part G, H, J, L or P of Schedule 1 imposes a requirement.

Controlled fittings

32 Where windows, roof windows, rooflights or doors are to be provided, reasonable provision would be the provision of draught-proofed units whose area-weighted average performance is no worse than given in Table 2. Column (a) applies to fittings provided as part of constructing an extension, column (b) to replacement fittings or new fittings installed in the existing **dwelling**.

33 The U-value or Window Energy Rating of a window, roof window or rooflight fittings can be taken as the value for either:

a. the standard configuration as set out in BR 443; or

b. the particular size and configuration of the actual fitting.

34 SAP 2005 Table 6e gives values for different window configurations that can be used in the absence of test data or calculated values.

Table 2 **Reasonable provision when working on controlled fittings**

Fitting	a) Standard for new fittings in extensions	(b) Standard for replacement fittings in an existing dwelling
Window, roof window and rooflight	U-value = 1.8W/m²·K or	U-value = 2.0W/m²·K or
	Window energy rating[15] = Band D; or	Window energy rating = Band E; or
	Centre-pane U-value = 1.2W/m²·K	Centre-pane U-value = 1.2W/m²·K
Doors with more than 50% of their internal face area glazed	2.2W/m²·K or centre-pane U-value = 1.2W/m²·K	2.2W/m²·K or centre-pane U-value = 1.2W/m²·K
Other doors	3.0W/m²·K	3.0W/m²·K

Controlled services

Heating and hot systems

35 Where the work involves the provision or extension of a heating or hot water system or part thereof, reasonable provision would be:

a. the installation of an appliance with an efficiency:

  i. not less than that recommended for its type in the Domestic Heating Compliance Guide[16]; and

  ii. where the appliance is the primary heating service, an efficiency which is not worse than two percentage points lower than that of the appliance being replaced. If the new appliance uses a different fuel, then the efficiency of the new appliance should be multiplied by the ratio of the CO_2 emission factor of the fuel used in the appliance being replaced to that of the fuel used in the new appliance before making this check. The CO_2 emission factors should be taken from Table 12 of SAP 2005. In the absence of specific information, the efficiency of the appliance being replaced may be taken from Table 4a or 4b of SAP 2005.

The aim is to discourage an existing appliance being replaced by a significantly less carbon efficient one. When fuel switching, if an old oil fired boiler with an efficiency of 72% is to be replaced by a dual solid fuel boiler with an efficiency of 65%, the equivalent efficiency of the dual solid fuel boiler would be 65% x (0.265/0.187) = 92.1%, and so the test in paragraph 35a)ii) would be satisfied. 0.265 and 0.187 kgCO₂/kWh are the emission factors for oil and dual fuel appliances respectively given in ADL1A.

b. the provision of controls that meet the minimum control requirements as given in the Domestic Heating Compliance Guide for the particular type of appliance and heat distribution system.

36 The heating and hot water system(s) should be commissioned so that at completion, the system(s) and their controls are left in working order and can operate efficiently for the purposes of the conservation of fuel and power. In order to demonstrate that the heating and hot water systems have been adequately commissioned, Regulation 20C states that:

20C.–(1) This regulation applies to building work in relation to which paragraph L1(b) of Schedule 1 imposes a requirement, but does not apply where the work consists only of work described in Schedule 2B.

(2) Where this regulation applies the person carrying out the work shall, for the purpose of ensuring compliance with paragraph L1(b) of Schedule 1, give to the local authority a notice confirming that the fixed building services have been commissioned in accordance with a procedure approved by the Secretary of State.

(3) The notice shall be given to the local authority:

(a) not later than the date on which the notice required by regulation 15(4) is required to be given; or

(b) where that regulation does not apply, not more than 30 days after completion of the work.

37 The procedure approved by the Secretary of State is set out in the Domestic Heating Compliance Guide.

38 The notice should include a declaration signed by someone suitably qualified to do so that the manufacturer's commissioning procedures have been completed satisfactorily.

One option would be to engage a member of an approved Competent Persons scheme.

---

[15] CE66 *Windows for new and existing housing,* EST, 2006.

[16] *Domestic Heating Compliance Guide,* NBS, 2006.

Insulation of pipes and ducts

39 As part of the provision or extension of a heating or hot water service, reasonable provision would be demonstrated by insulating pipes ducts and vessels to standards that are not worse than those set out in the Domestic Heating Compliance Guide.

The TIMSA Guide[17] explains the derivation of the performance standards and how they can be interpreted in practice.

Mechanical ventilation

40 Where the work involves the provision of a mechanical ventilation system or part thereof, reasonable provision would be to install systems no worse than those described in GPG 268[18] which also have specific fan powers and heat recovery efficiency not worse than those in Table 3.

Table 3 Limits on design flexibility for mechanical ventilation systems

System type	Performance
Specific Fan Power (SFP) for continuous supply only and continuous extract only	0.8 litre/s.W
SFP for balanced systems	2.0 litre/s.W
Heat recover efficiency	66%

41 Mechanical ventilation systems must satisfy the requirements in Part F.

Mechanical cooling

42 Where the work involves the provision of a fixed household air conditioner, reasonable provision would be to provide a unit having an energy efficiency classification equal to or better than class C in Schedule 3 of the labelling scheme adopted under The Energy Information (Household Air Conditioners) (No. 2) Regulations 2005[19].

Fixed internal lighting

43 Reasonable provision should be made for *dwelling* occupiers to obtain the benefits of efficient electric lighting whenever

a. a *dwelling* is extended; or

b. a new *dwelling* is created from a material change of use; or

c. an existing lighting system is being replaced as part of re-wiring works.

The re-wiring works must comply with Part P.

44 A way of showing compliance would be to provide lighting fittings (including lamp, control gear and an appropriate housing, reflector, shade or diffuser or other device for controlling the output light) that only take lamps having a luminous efficacy greater than 40 lumens per circuit-Watt.

Circuit-Watts means the power consumed in lighting circuits by lamps and their associated control gear and power factor correction equipment.

Fluorescent and compact fluorescent lighting fittings would meet this standard. Lighting fittings for GLS tungsten lamps with bayonet cap or Edison screw bases, or tungsten halogen lamps would not.

45 Reasonable provision would be to provide in the areas affected by the building work, fixed energy efficient light fittings that number not less than the greater of:

a. one per 25m² of *dwelling* floor area (excluding garages) or part thereof; or

b. one per four fixed lighting fittings.

*This assessment should be based on the extension, the newly created **dwelling** or the area served by the lighting system as appropriate to the particular case.*

Installing mains frequency fluorescent lighting in garages may cause dangers through stroboscopic interaction with vehicle engine parts or machine tools. Fluorescent lamps with high frequency electronic ballasts substantially reduce this risk.

46 A light fitting may contain one or more lamps.

47 Lighting fittings in less used areas like cupboards and other storage areas would not count towards the total. GIL 20[20] gives guidance on identifying suitable locations for energy efficient luminaires. In some cases, it may be more appropriate to install the energy efficient light fitting in a location that is not part of the building work, e.g. to replace the fitting on the landing when creating a new bedroom through a loft conversion.

Fixed external lighting

*Fixed external lighting means lighting fixed to an external surface of the **dwelling** supplied from the occupier's electrical system. It excludes the lighting in common areas in blocks of flats and other access-way lighting provided communally.*

48 When providing fixed external lighting, reasonable provision should be made to enable effective control and/or the use of efficient lamps such that:

a. EITHER: Lamp capacity does not exceed 150 Watts per light fitting and the lighting automatically switches off:

   i. When there is enough daylight; and

   ii. When it is not required at night; or

b. the lighting fittings have sockets that can only be used with lamps having an efficacy greater than 40 lumens per circuit-Watt.

Compact fluorescent lamp types would meet the standard in (b), but GLS tungsten lamps with bayonet cap or Edison screw bases, or tungsten halogen lamps would not.

[17] *HVAC Guidance for Achieving Compliance with Part L of the Building Regulations*, TIMSA 2006.

[18] GPG268 *Energy efficient ventilation in dwellings – guide for specifiers*, EST, 2006.

[19] Statutory Instrument SI 2005/1726, the Energy Information (Household Air Conditioners) (No. 2) Regulations 2005.

[20] GIL20, *Low energy domestic lighting*, EST, 2006.

Section 2: Guidance on thermal elements

49 New *thermal elements* must comply with requirement L1(a)(i). Work on existing elements is covered by regulation 4A which states:

> **4A.**–(1) Where a person intends to renovate a thermal element, such work shall be carried out as is necessary to ensure that the whole thermal element complies with the requirements of paragraph L1(a)(i) of Schedule 1.
>
> (2) Where a thermal element is replaced, the new thermal element shall comply with the requirements of paragraph L1(a)(i) of Schedule 1.

THE PROVISION OF THERMAL ELEMENTS

U-values

50 Reasonable provision for newly constructed *thermal elements* such as those constructed as part of an extension would be to meet the standards set out in column (a) of Table 4. In addition, no individual element should have a U-value worse than those set out in column (b) of Table 1.

51 Reasonable provision for those *thermal elements* constructed as replacements for existing elements would be to meet the standards set out in column (b) of Table 4. In addition, no part of a *thermal element* should have a U-value worse than those set out in column (b) of Table 1.

Continuity of insulation and airtightness

52 The building fabric should be constructed so that there are no reasonably avoidable thermal bridges in the insulation layers caused by gaps within the various elements, at the joints between elements and at the edges of elements such as those around window and door openings. Reasonable provision should also be made to reduce unwanted air leakage through the new envelope parts.

53 A suitable approach to showing the requirement has been achieved would be to submit a report signed by a suitably qualified person confirming that appropriate design details and building techniques have been specified, and that the work has been carried out in ways that can be expected to achieve reasonable conformity with the specifications. Reasonable provision would be to:

a. adopt design details such as those set out in the TSO Robust Details catalogue[21]; or

A list of additional approved details may be provided in due course.

b. to demonstrate that the specified details deliver an equivalent level of performance using the guidance in BRE IP 1/06[22].

Table 4 Standards for thermal elements W/m²·K

Element[1]	(a) Standard for new thermal elements in an extension	(b) Standard for replacement thermal elements in an existing dwelling
Wall	0.30	0.35[2]
Pitched roof – insulation at ceiling level	0.16	0.16
Pitched roof – insulation at rafter level	0.20	0.20
Flat roof or roof with integral insulation	0.20	0.25
Floors	0.22[3]	0.25[3]

Notes:

1. Roof includes the roof parts of dormer windows and wall refers to the wall parts (cheeks) of dormer windows.

2. A lesser provision may be appropriate where meeting such a standard would result in a reduction of more than 5% in the internal floor area of the room bounded by the wall.

3. A lesser provision may be appropriate where meeting such a standard would create significant problems in relation to adjoining floor levels. The U-value of the floor of an extension can be calculated using the exposed perimeter and floor area of the whole enlarged *dwelling*.

---

[21] *Limiting thermal bridging and air leakage: Robust construction details for dwellings and similar buildings*, Amendment 1, TSO, 2002. See www.est.org.uk.

[22] IP 1/06 *Assessing the effects of thermal bridging at junctions and around openings in the external elements of buildings*, BRE 2006.

RENOVATION OF THERMAL ELEMENTS

54 Where a *thermal element* is being renovated reasonable provision in most cases would be to achieve the standard set out in column (b) of Table 5. Where the works apply to less than 25% of the surface area however reasonable provision could be to do nothing to improve energy performance.

55 If such an upgrade is not technically or functionally feasible or would not achieve a *simple payback* of 15 years or less, the element should be upgraded to the best standard that is technically and functionally feasible and which can be achieved within a *simple payback* of no greater than 15 years. Guidance on this approach is given in Appendix A.

RETAINED THERMAL ELEMENTS

56 Part L applies to retained *thermal elements* in the following circumstances:

a. where an existing *thermal element* is part of a building subject to a material change of use;

b. where an existing element is to become part of the thermal envelope and is to be upgraded.

57 Reasonable provision would be to upgrade those *thermal elements* whose U-value is worse than the threshold value in column (a) of Table 5 to achieve the U-value given in column (b) of Table 5 provided this is technically, functionally and economically feasible. A reasonable test of economic feasibility is to achieve a *simple payback* of 15 years or less. Where the standard given in column (b) is not technically, functionally or economically feasible, then the element should be upgraded to the best standard that is technically and functionally feasible and delivers a simple payback period of 15 years or less.

Examples of where lesser provision than column (b) might apply are where the thickness of the additional insulation might reduce usable floor area by more than 5% or create difficulties with adjoining floor levels, or where the weight of the additional insulation might not be supported by the existing structural frame.

Table 5 **Upgrading retained thermal elements**

Element	(a) Threshold value W/m²·K	(b) Improved value W/m²·K
Cavity wall*	0.70	0.55
Other wall type	0.70	0.35
Floor	0.70	0.25
Pitched roof – insulation at ceiling level	0.35	0.16
Pitched roof – insulation between rafters	0.35	0.20
Flat roof or roof with integral insulation	0.35	0.25

* This only applies in the case of a wall suitable for the installation of cavity insulation. Where this is not the case it should be treated as for 'other wall type.'

Section 3: Providing information

58 On completion of the work, in accordance with requirement L1(c), the owner of the **dwelling** should be provided with sufficient information about the building, the fixed building services and their maintenance requirements so that the building can be operated in such a manner as to use no more fuel and power than is reasonable in the circumstances.

59 A way of complying would be to provide a suitable set of operating and maintenance instructions aimed at achieving economy in the use of fuel in terms that householders can understand in a durable format that can be kept and referred to over the service life of the system(s). The instructions should be directly related to the particular system(s) installed as part of the work that has been carried out.

The aim is that this information could eventually form part of the Home Information Pack.

60 Without prejudice to the need to comply with health and safety requirements, the instructions should explain to the occupier of the **dwelling** how to operate the system(s) efficiently. This should include

a. the making of adjustments to the timing and temperature control settings and

b. what routine maintenance is needed to enable operating efficiency to be maintained at a reasonable level through the service life/lives of the system(s).

61 Where a new **dwelling** is created by a material change of use (see paragraphs 25 to 28), an energy rating shall be prepared and fixed in a conspicuous place in the **dwelling** as required by Regulation 16, which states that:

16.–(1) This regulation applies where a new dwelling is created by building work or by a material change of use in connection with which building work is carried out.

(2) Where this regulation applies, the person carrying out the building work shall calculate the energy rating of the dwelling by means of a procedure approved by the Secretary of State and give notice of that rating to the local authority.

(3) The notice referred to in paragraph (2) shall be given not later than the date on which the notice required by paragraph (4) of regulation 15 is given, and, where a new dwelling is created by the erection of a building, it shall be given at least five days before occupation of the dwelling.

(4) Where this regulation applies, subject to paragraphs (6) and (7), the person carrying out the building work shall affix, as soon as practicable, in a conspicuous place in the dwelling, a notice stating the energy rating of the dwelling.

(5) The notice referred to in paragraph (4) shall be affixed not later than the date on which the notice required by paragraph (4) of regulation 15 is given, and, where a new dwelling is created by the erection of a building, it shall be affixed not later than five days before occupation of the dwelling.

(6) Subject to paragraph (7), if, on the date the dwelling is first occupied as a residence, no notice has been affixed in the dwelling in accordance with paragraph (4), the person carrying out the building work shall, not later than the date on which the notice required by paragraph (4) of regulation 15 is given, give to the occupier of the dwelling a notice stating the energy rating of the dwelling calculated in accordance with paragraph (2).

(7) Paragraphs (4) and (6) shall not apply in a case where the person carrying out the work intends to occupy, or occupies, the dwelling as a residence.

62 The approved calculation procedure is SAP 2005 as announced in ODPM Circular 03/2006.

63 Guidance on the preparation of the notices is given in DTLR Circular 3/2001.

Section 4: Definitions

64 For the purpose of this Approved Document, the following definitions apply.

65 *BCB* means Building Control Body: a local authority or an approved inspector.

66 A ***conservatory*** is an extension to a building which:

a. has not less than three quarters of its roof area and not less than one half of its external wall area made from translucent material and

b. is thermally separated from the ***dwelling*** by walls, windows and doors with the same U-value and draught-stripping provisions as provided elsewhere in the ***dwelling***.

67 *Dwelling* means a self-contained unit designed to be used separately to accommodate a single household.

68 *Energy efficiency requirements* means the requirements of Regulations 4A, 17C and 17D and Part L of Schedule 1.

69 *Fixed building services* means any part of, or any controls associated with:

a. fixed internal or external lighting systems but does not include emergency escape lighting and specialist process lighting; or

b. fixed systems for heating, hot water service systems, air conditioning or mechanical ventilation.

70 *Renovation* in relation to a thermal element means the provision of a new layer in the thermal element or the replacement of an existing layer, but excludes decorative finishes, and 'renovate' shall be construed accordingly.

71 *Room for residential purposes* means a room, or a suite of rooms, which is not a dwelling-house or a flat and which is used by one or more persons to live and sleep and includes a room in a hostel, a hotel, a boarding house, a hall of residence or a residential home, whether or not the room is separated from or arranged in a cluster group with other rooms, but does not include a room in a hospital, or other similar establishment, used for patient accommodation and, for the purposes of this definition, a 'cluster' is a group of rooms for residential purposes which is:

a. separated from the rest of the building in which it is situated by a door which is designed to be locked; and

b. not designed to be occupied by a single household.

72 *Simple payback* means the amount of time it will take to recover the initial investment through energy savings, and is calculated by dividing the marginal additional cost of implementing an energy efficiency measure by the value of the annual energy savings achieved by that measure taking no account of VAT.

a. The marginal additional cost is the additional cost (materials and labour) of incorporating (e.g.) **additional** insulation, not the whole cost of the work.

b. The cost of implementing the measure should be based on prices current at the date the proposals are made known to the ***BCB*** and be confirmed in a report signed by a suitably qualified person.

c. The annual energy savings should be estimated using SAP 2005 or SBEM[23].

d. For the purposes of this Approved Document, the following energy prices that were current in 2005 should be used when evaluating the value of the annual energy savings:

   i. Mains gas – 1.63 p/kWh

   ii. Electricity – 3.65 p/kWh

This is a weighted combination at peak and off peak tariffs.

   iii. Heating oil – 2.17 p/kWh

   iv. LPG – 3.71 p/kWh.

For example if the cost of implementing a measure was £430 and the value of the annual energy savings was £38/year, the simple payback would be (430/38) = 11.3 years.

Energy prices are increasing significantly so dwelling owners may wish to use higher values such as those prevailing when they apply for Building Regulations approval.

73 Thermal ***element*** is defined in Regulation 2A as follows.

(2A) In these Regulations 'thermal element' means a wall, floor or roof (but does not include windows, doors, roof windows or roof-lights) which separates a thermally conditioned part of the building ('the conditioned space') from:

a. the external environment (including the ground); or

b. in the case of floors and walls, another part of the building which is:

   i. unconditioned;

   ii. an extension falling within class VII of Schedule 2; or

   iii. where this paragraph applies, conditioned to a different temperature,

and includes all parts of the element between the surface bounding the conditioned space and the external environment or other part of the building as the case may be.

(2B) Paragraph (2A)(b)(iii) only applies to a building which is not a dwelling, where the other part of the building is used for a purpose which is not similar or identical to the purpose for which the conditioned space is used.

---

[23] Simplified Building Energy Model (SBEM) user manual and Calculation Tool, available at www.odpm.gov.uk

Appendix A: Work to thermal elements

1 Where the work involves the *renovation* of a *thermal element*, an opportunity exists for cost-effective insulation improvements to be undertaken at marginal additional cost. This appendix provides guidance on the cost effectiveness of insulation measures when undertaking various types of work on a *thermal element*.

2 Table A1 sets out the circumstances and the level of performance that would be considered reasonable provision in ordinary circumstances. When dealing with existing *dwellings* some flexibility in the application of standards is necessary to ensure that the context of each scheme can be taken into account while securing, as far as possible, the reasonable improvement. The final column in Table A1 provides guidance on a number of specific issues that may need to be considered in determining an appropriate course of action. As part of this flexible approach, it will be necessary to take into account technical risk and practicality in relation to the *dwelling* under consideration and the possible impacts on any adjoining building. In general the proposed works should take account of:

a. the other parts of Schedule 1; and

b. the general guidance on technical risk relating to insulation improvements contained in BR 262; and

c. if the existing building has historic value, the guidance produced by English Heritage.

Where it is not reasonable in the context of the scheme to achieve the performance set out in Table A1, the level of performance achieved should be as close to this as practically possible.

3 Table A1 incorporates, in outline form, examples of construction that would achieve the proposed performance, but designers are free to use any appropriate construction that satisfies the energy performance standard, so long as they do not compromise performance with respect to any other part of the regulations.

GENERAL GUIDANCE

4 This section lists general guidance documents that provide advice on the *renovation* options available and their application. The listing of any guide, British Standard or other document does not indicate that the guidance contained is approved or applicable to any particular scheme. It is for the applicant and his or her advisors to assess the applicability of the guidance in the context of a particular application. Responsibility for the guidance contained in the documents listed rests with the authors and authoring organisations concerned.

5 In a number of documents (particularly those produced by the Energy Saving Trust's Energy Efficiency Best Practice in Housing programme) a recommended thermal performance is stated in the form of a U-value for different elements and forms of construction. The inclusion of such a performance value in any guidance document in this Appendix does not constitute a performance limit or target for the purposes of this Approved Document. In all cases the relevant target U-values are those contained in Table A1.

General guidance

Stirling, C. (2002) Thermal insulation: Avoiding Risks, Building Research Establishment report BR 262, Watford, Construction Research Communications Ltd.

EST (2004) Energy efficient refurbishment of existing housing, Good Practice guide 15, Energy Efficiency Best Practice in Housing, London, Energy Saving Trust.

EST (2004) Refurbishing Cavity Walled Dwellings, CE 57, Energy Efficiency Best Practice in Housing, London, Energy Saving Trust.

EST (2004) Refurbishing Dwellings with Solid Walls, CE 58, Energy Efficiency Best Practice in Housing, London, Energy Saving Trust.

EST (2004) Refurbishing Timber-Framed Dwellings, CE 59, Energy Efficiency Best Practice in Housing, London, Energy Saving Trust.

EST (2005) Advanced Insulation in housing Refurbishment, CE 97, Energy Efficiency Best Practice in Housing, London, Energy Saving Trust.

Roofs

EST (2002) Refurbishment Site Guidance for Solid-Walled Houses – Roofs, GPG 296, Energy Efficiency Best Practice in Housing, London, Energy Saving Trust.

Stirling (2000) Insulating roofs at rafter level: sarking insulation, Good Building Guide 37, Watford, Building Research Establishment.

Code of practice for loft insulation: National Insulation Association.

Walls

EST (2000) External Insulation Systems for Walls of Dwellings, GPG 293, Energy Efficiency Best Practice in Housing, London, Energy Saving Trust.

EST (2000) Refurbishment Site Guidance for Solid-Walled Houses – Walls, GPG 297, Energy Efficiency Best Practice in Housing, London, Energy Saving Trust.

EST (2003) Internal Wall insulation in Existing housing, GPG 138, Energy Efficiency Best Practice in Housing, London, Energy Saving Trust.

Floors

EST (2002) Refurbishment Site Guidance for Solid-Walled Houses – Ground Floors, GPG 294, Energy Efficiency Best Practice in Housing, London, Energy Saving Trust.

International, European and British Standards

BS 5250:2002 Code of practice for the control of condensation in buildings.

BS EN ISO 13788:2001 Hygrothermal performance of building components and building elements. Internal surface temperature to avoid critical surface humidity and interstitial condensation. Calculation methods.

BS 6229:2003, Flat roofs with continuously supported coverings – Code of practice.

BS 5803-5:1985, Thermal insulation for use in pitched roof spaces in dwellings. Specification for installation of man-made mineral fibre and cellulose fibre insulation. Amended 1999 incorporating amendment no.1 1994.

Table A1 Cost-effective U-value targets when undertaking renovation works to *thermal elements*

Proposed works	Target U-value (W/m²·K)	Typical construction	Comments (reasonableness, practicability and cost-effectiveness)
Pitched roof constructions			
Renewal of roof covering – No living accommodation in the roof void – existing insulation (if any) at ceiling level. No existing insulation, existing insulation less than 50mm, in poor condition, and/or likely to be significantly disturbed or removed as part of the planned work	0.16	Provide loft insulation – 250mm mineral fibre or cellulose fibre as quilt laid between and across ceiling joists or loose fill or equivalent. This may be inappropriate if the loft is already boarded out and the boarding is not to be removed as part of the work	Assess condensation risk in roof space and make appropriate provision in accordance with the requirements of Part C relating to the control of condensation. Additional provision may be required to provide access to and insulation of services in the roof void
Renewal of roof covering – Existing insulation in good condition and will not be significantly disturbed by proposed works. Existing insulation thickness 50mm or more but less than 100mm	0.20	Top-up loft insulation to at least 200mm mineral fibre or cellulose fibre as quilt laid between and across ceiling joists or loose fill or equivalent	Assess condensation risk in roof space and make appropriate provision in line with the requirements of Part C relating to the control of condensation. Additional provision may be required to provide insulation and access to services in the roof void Where the loft is already boarded out and the boarding is not to be removed as part of the work, the practicality of insulation works would need to be considered
Renewal of the ceiling to cold loft space. Existing insulation at ceiling level removed as part of the works	0.16	Provide loft insulation – 250mm mineral fibre or cellulose fibre as quilt laid between and across ceiling joists or loose fill or equivalent	Assess condensation risk in roof space and make appropriate provision in accordance with the requirements of Part C relating to the control of condensation. Additional provision may be required to provide insulation and access to services in the roof void Where the loft is already boarded out and the boarding is not to be removed as part of the work, insulation can be installed from the underside but the target U-value may not be achievable

Table A1 Cost-effective U-value targets when undertaking renovation works to *thermal elements*

Proposed works	Target U-value (W/m²·K)	Typical construction	Comments (reasonableness, practicability and cost-effectiveness)
Renewal of roof covering – Living accommodation in roof space (room-in-the-roof type arrangement), with or without dormer windows	0.20	Cold structure – Insulation (thickness dependent on material) placed between and below rafters Warm structure – Insulation placed between and above rafters	Assess condensation risk (particularly interstitial condensation), and make appropriate provision in accordance with the requirements of Part C relating to the control of condensation (Clause 8.4 of BS 5250:2002 and BS EN ISO, 13788:2001) Practical considerations with respect to an increase in structural thickness (particularly in terraced dwellings) may necessitate a lower performance target
Dormer window constructions			
Renewal of cladding to side walls	0.35	Insulation (thickness dependent on material) placed between and/or fixed to outside of wall studs. Or fully external to existing structure depending on construction	Assess condensation risk and make appropriate provision in accordance with the requirements of Part C
Renewal of roof covering	–	Follow guidance on improvement to pitched or flat roofs as appropriate	Assess condensation risk and make appropriate provision in accordance with the requirements of Part C
Flat roof constructions			
Renewal of roof covering – Existing insulation, if any, less than 100mm, mineral fibre (or equivalent resistance), or in poor condition and likely to be significantly disturbed or removed as part of the planned work	0.25	Insulation placed between and over joists as required to achieve the target U-value – Warm structure	Assess condensation risk and make appropriate provision in accordance with the requirements of Part C. Also see BS 6229:2003 for design guidance
Renewal of the ceiling to flat roof area. Existing insulation removed as part of the works	0.25	Insulation placed between and to underside of joists to achieve target U-value	Assess condensation risk and make appropriate provision in accordance with the requirements of Part C. Also see BS 6229:2003 for design guidance Where ceiling height would be adversely affected, a lower performance target may be appropriate

Table A1 Cost-effective U-value targets when undertaking renovation works to thermal elements

Proposed works	Target U-value (W/m²·K)	Typical construction	Comments (reasonableness, practicability and cost-effectiveness)
Solid wall constructions			
Renewal of internal finish to external wall or applying a finish for the first time	0.35	Dry-lining to inner face of wall – insulation between studs fixed to wall to achieve target U-value – thickness dependent on insulation and stud material used Insulated wall board fixed to internal wall surface to achieve the required U-value – thickness dependent on material used	Assess the impact on internal floor area. In general it would be reasonable to accept a reduction of no more than 5% of the area of a room. However, the use of the room and the space requirements for movement and arrangements of fixtures, fittings and furniture should be assessed In situations where acoustic attenuation issues are particularly important (e.g. where insulation is returned at party walls) a less demanding U-value may be more appropriate. In such cases, the U-value target may have to be increased to 0.35 or above depending on the circumstances Assess condensation and other moisture risks and make appropriate provision in accordance with the requirements of Part C. This will usually require the provision of a vapour control and damp protection to components. Guidance on the risks involved is provided in Sterling (2002) and, on the technical options, in EST (2003)
Renewal of finish or cladding to external wall area or elevation (render or other cladding) or applying a finish or cladding for the first time	0.35	External insulation system with rendered finish or cladding to give required U-value	Assess technical risk and impact of increased wall thickness on adjoining buildings
Cavity wall constructions			
Replace wall ties to at least one elevation	0.55	Include cavity wall insulation	Assess suitability of cavity for full fill insulation in accordance with requirements of Part C
Ground floor constructions			
Renovation of a solid or suspended floor involving the replacement of screed or a timber floor deck	See comment	Solid floor – replace screed with an insulated floor deck to maintain existing floor level Suspended timber floor – fit insulation between floor joists prior to replacement of floor deck	The cost-effectiveness of floor insulation is complicated by the impact of the size and shape of the floor (Perimeter/Area ratio). In many cases existing uninsulated floor U-values are already relatively low when compared with wall and roof U-values. Where the existing floor U-value is greater than 0.70W/m²·K, then the addition of insulation is likely to be cost-effective. Analysis shows that the cost–benefit curve for the thickness of added insulation is very flat, and so a target U-value of 0.25W/m²·K is appropriate subject to other technical constraints (adjoining floor levels, etc.)

Documents referred to

BRE
www.bre.co.uk

BR 262 *Thermal insulation: avoiding risks*, 2001. ISBN 1 86081 515 4

BRE Report BR 443 *Conventions for U-value calculations*, 2006.
(Available at www.bre.co.uk/uvalues.)

Information Paper IP1/06 *Assessing the effects of thermal bridging at junctions and around openings in the external elements of buildings*, 2006. ISBN 1 86081 904 4

Simplified Building Energy Model (SBEM) user manual and Calculation Tool.
(Available from www.odpm.gov.uk.)

Department of the Environment, Food and Rural Affairs (Defra)
www.defra.gov.uk

The Government's Standard Assessment Procedure for energy rating of dwellings, SAP 2005.
(Available at www.bre.co.uk/sap2005.)

Department of Transport, Local Government and the Regions (DTLR)

Limiting thermal bridging and air leakage: Robust construction details for dwellings and similar buildings, Amendment 1. Published by TSO, 2002. ISBN 0 11753 631 8
(or download from Energy Saving Trust (EST) website on http://portal.est.org.uk/ housingbuildings/calculators/robustdetails/.)

Energy Saving Trust (EST)
www.est.org.uk

CE66, *Windows for new and existing housing*, 2006.

GPG268, *Energy efficient ventilation in dwellings – a guide for specifiers*, 2006.

GIL 20, *Low energy domestic lighting*, 2006.

English Heritage
www.english-heritage.org.uk

Building Regulations and Historic Buildings, 2002 (revised 2004).

Health and Safety Executive (HSE)
www.hse.gov.uk

L24 *Workplace Health, Safety and Welfare: Workplace (Health, Safety and Welfare) Regulations 1992, Approved Code of Practice and Guidance, The Health and Safety Commission*, 1992. ISBN 0 71760 413 6

NBS (on behalf of ODPM)
www.thebuildingregs.com

Domestic Heating Compliance Guide, 2006.
ISBN 1 85946 225 1

Thermal Insulation Manufacturers and Suppliers Association (TIMSA)
www.timsa.org.uk

HVAC Guidance For Achieving Compliance With Part L of the Building Regulations, 2006.

Legislation

SI 1991/1620 Construction Products Regulations 1991.

SI 1992/2372 Electromagnetic Compatibility Regulations 1992.

SI 1994/3051 Construction Products (Amendment) Regulations 1994.

SI 1994/3080 Electromagnetic Compatibility (Amendment) Regulations 1994.

SI 1994/3260 Electrical Equipment (Safety) Regulations 1994.

SI 2001/3335 Building (Amendment) Regulations 2001.

SI 2005/1726 Energy Information (Household Air Conditioners) (No. 2) Regulations 2005.

SI 2006/652 Building And Approved Inspectors (Amendment) Regulations 2006.

Standards referred to

BS 8206-2:1992 Lighting for buildings. Code of
practice for daylighting.